How Does a Farmer Use Science?

Written and Designed by Alix Wood

We like to visit Peter and Penny's farm.

Peter and Penny grow food and raise animals on the farm.

They use **science** to help them.

Corn plants

Penny knows what nutrients to feed the hens so they lay eggs for our breakfast.

Chicken food

Some hens share a pen with a male chicken, called a cockerel.

The eggs of these hens are fertilised by the cockerel.

Hen
Cockerel

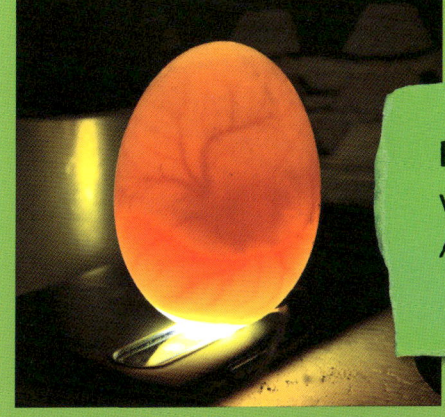

Penny checks a fertilised egg with a special light machine. A chick is growing inside!

Hen sitting on eggs

After 21 days under a warm hen, a fluffy chick hatches from each fertilised egg.

Chick

Farmers have to know about animal science to take care of their animals.

The first milk a nanny goat feeds her newborn kids is full of antibodies that fight diseases.

Let's say it! "AN-tee-boh-deez"

Nanny goat

Goat kid

One of the kids on the farm is smaller than the others.

I've named the baby goat Henry.

An extra bottle of his mother's milk will help him to grow.

Peter washes Henry's bottles with boiling water to kill any germs.

As the goat kids get older, they don't need milk. Peter turns the spare milk into cheese.

Get ready for some BIG science!

1 Peter slowly heats some milk in a stainless steel pan.

2 He checks the milk's temperature with a thermometer.

Goats' milk

Thermometer

Stainless steel pan

3 When the temperature is just right, he stirs in some lemon juice.

Lumpy milk

4 As the milk cools, the lemon juice makes the milk get lumpy.

Cheesecloth

5 The lumps are wrapped in cheesecloth and hung over a pot.

The leftover milk drains away.

Mould

6 Finally, Peter adds salt to the cheese and then sets it in round moulds.

Cheese

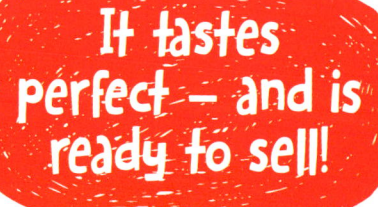

It tastes perfect – and is ready to sell!

Penny must dig over a field so she can sow some corn seeds.

She chooses a plough and tractor to use.

A farmer must know the weight of a plough.

Then the farmer chooses a tractor with enough power to pull that plough.

Plough

Tractor

Ploughed soil

Modern tractors have computers inside.

A screen shows the farmer a picture of the field.

The computer helps the farmer plough and sow seeds in straight lines.

Picture of field

Computer

Some computers can help steer the tractor!

Peter built a beehive and put it in the apple orchard.

Lots of honeybees live in the hive.

Beehive

The bees collect sweet nectar and pollen from the apple blossom. They make honey with the nectar.

The farmers let the hedgerows around their fields grow all summer.

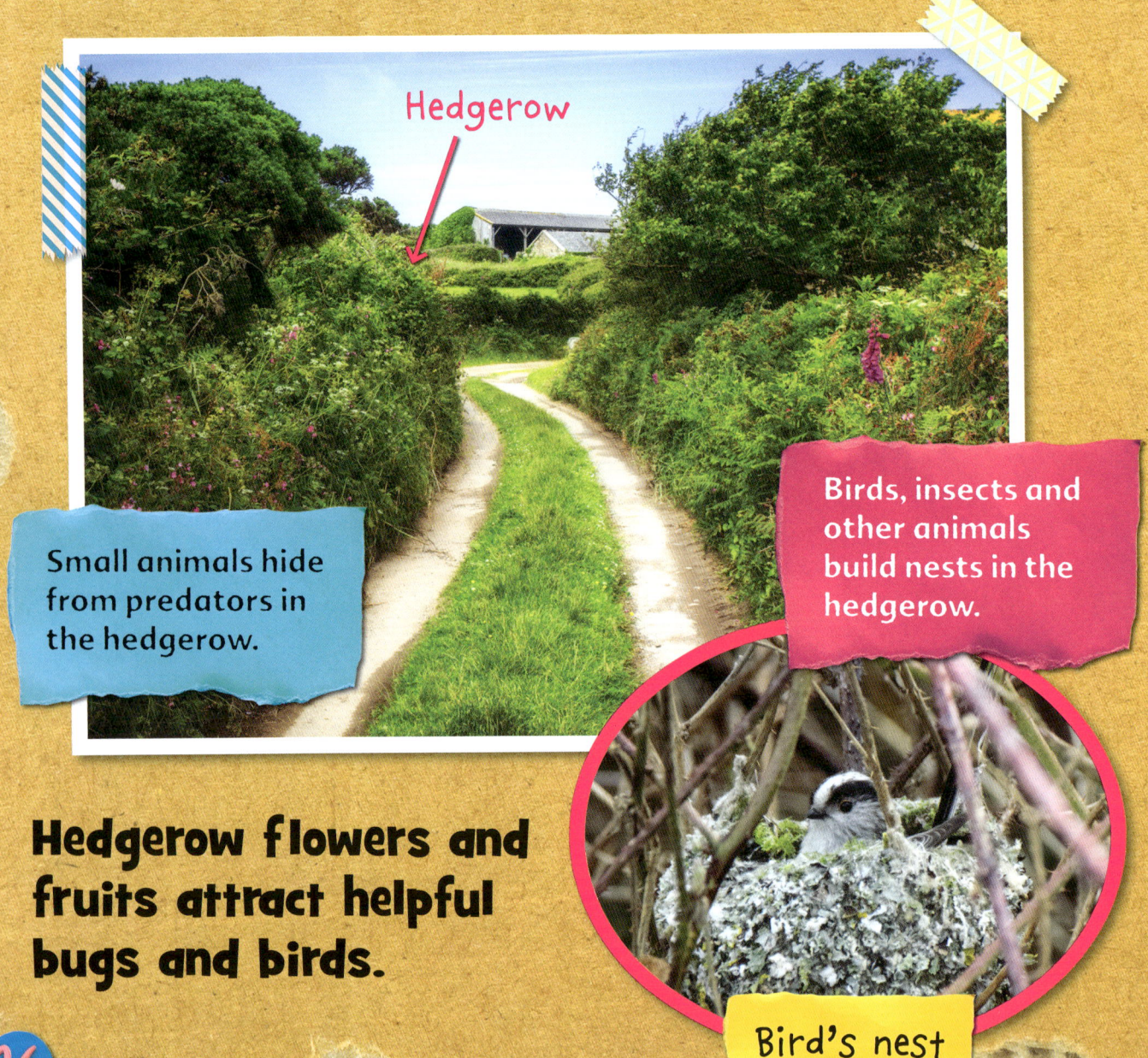

Hedgerow

Small animals hide from predators in the hedgerow.

Birds, insects and other animals build nests in the hedgerow.

Bird's nest

Hedgerow flowers and fruits attract helpful bugs and birds.

Onion flower

Butterfly

Bees and butterflies help to pollinate the crops.

Birds help by eating bugs that can harm the crops.

Having lots of different plants and wild animals is called biodiversity. It's good for the countryside!

Let's say it!
"bi-oh-die-VUR-suh-tee"

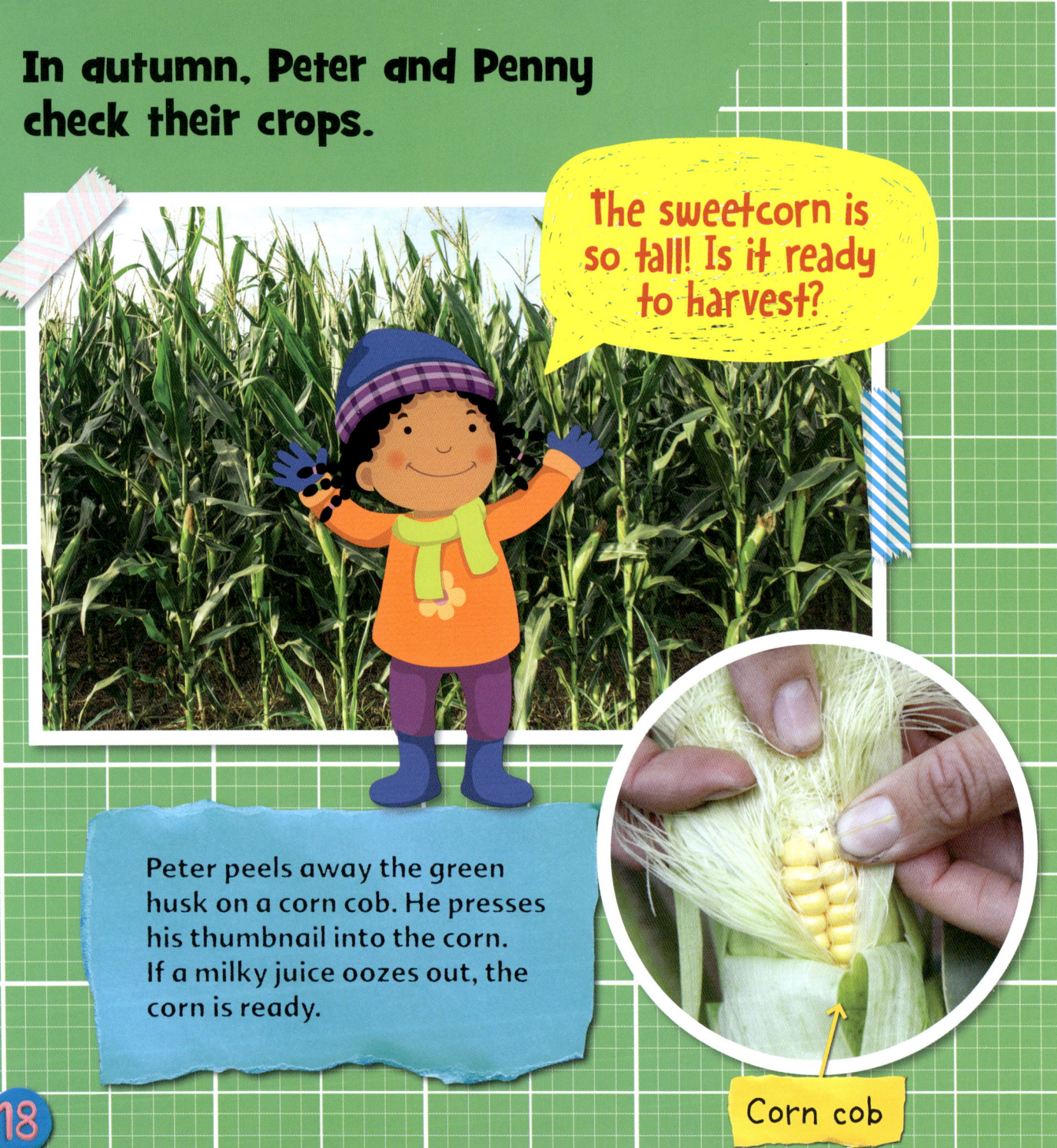

The farm's sheep like the solar panels, too.

Solar panels

They hide under the panels to shelter from the Sun.

A tractor pulling a mowing machine would never fit in there.

The sheep help the farmers by mowing the grass under the solar panels.

Today is the BIG village show and Henry needs to get ready!

The best goat at the show will win a prize. Penny trims Henry's hooves...

Caring for an animal's feet stops it getting foot diseases.

Hoof-cutting tool

...she washes and brushes Henry's coat.

This stops ticks and fleas that can make animals sick.

Science helps Peter and Penny when they breed goats.

Henry's dad

Henry's mum

They picked the healthiest, best-looking mum and dad goat.

Together, they made healthy, handsome Henry.

Science worked! Henry won first prize.

Now we know how farmers use science. Good work, little scientists!

My Science Words

antibodies
Tiny parts in an animal or person's blood that help to protect them from germs and viruses.

fertilised
Ready to grow into a new living thing.

nutrients
Substances, such as vitamins, that a living thing needs to help it live and grow.

pollinate
To carry pollen from one flower to another. This helps the flowers make seeds.